THE BAKERY FACTORY

THE BAKERY FACTORY

Who Puts the Bread on Your Table

Text and photographs by

AYLETTE JENNESS

Thomas Y. Crowell Company · New York

I would like to thank the people I talked to, photographed, and learned from at Dorothy Muriel's Bakery, especially Bernadette Nicotra, Abram Blanken, and Walter Rajewski. Their friendliness, trust, and willingness to share their knowledge made this book possible.

I also want to thank friends and family who worked on the book: Corinna and Eve Bowles, Danny and Matt Kobin, Mark Field, Jessica Trefonides, and Bonnie Rottier.

Lastly, special gratitude to Matthew Aldrich for his fine drawings, Sam Bowles for help, help, and more help, and Evan Jenness, for coming along.

 Published simultaneously in Canada by Fitzhenry & Whiteside Limited, Toronto.
Manufactured in the United States of America

Library of Congress Cataloging in Publication Data
Jenness, Aylette. The bakery factory.
Includes index.
SUMMARY: Text and photos introduce a large commercial bakery, its machines, products, and employees.
1. Bakers and bakeries—Juv. lit. [1. Bakers and bakeries] I. Title.
TX763.J47 664'.752 77-8094 ISBN 0-690-03805-4
0-690-01338-8 (lib. bdg.)

1 2 3 4 5 6 7 8 9 10

To the workers
at Dorothy Muriel's Bakery
and to my family,
old and new

By Aylette Jenness

Along the Niger River: An African Way of Life
The Bakery Factory: Who Puts the Bread on Your Table

By Aylette Jenness and Lisa W. Kroeber

A Life of Their Own: An Indian Family in Latin America

CONTENTS

I.

LOOKING AROUND

Have you ever wondered where the things you use and wear every day come from? You probably know the stores in your town where they were bought, but where did they come from before *that*?

What about your sneakers? Were they made here in the United States or in some other country? Do you think they were made in a factory where thousands of pairs of tennis shoes are turned out every day?

What about your softball? Someone must have sewn the pieces of leather together. Who did it? Someone old or young? A woman or a man? Did that person like making the ball? Did she or he make the ball from start to finish? Was it hard to do? What's it like to work in a softball factory?

What about your pencil? How *are* pencils made, anyway? I thought there must be machines that drill long tiny holes in pencil-shaped pieces of wood, and other machines that push in the sticks of graphite. Then I read about pencil making in the encyclopedia, and I found out I was all wrong. Machines cut long grooves in thin pieces of wood. Then the graphite sticks are laid in the grooves. Another grooved piece of wood, like the first one, is glued on top. That "sandwich" is trimmed into a pencil shape, then dipped into a vat of paint.

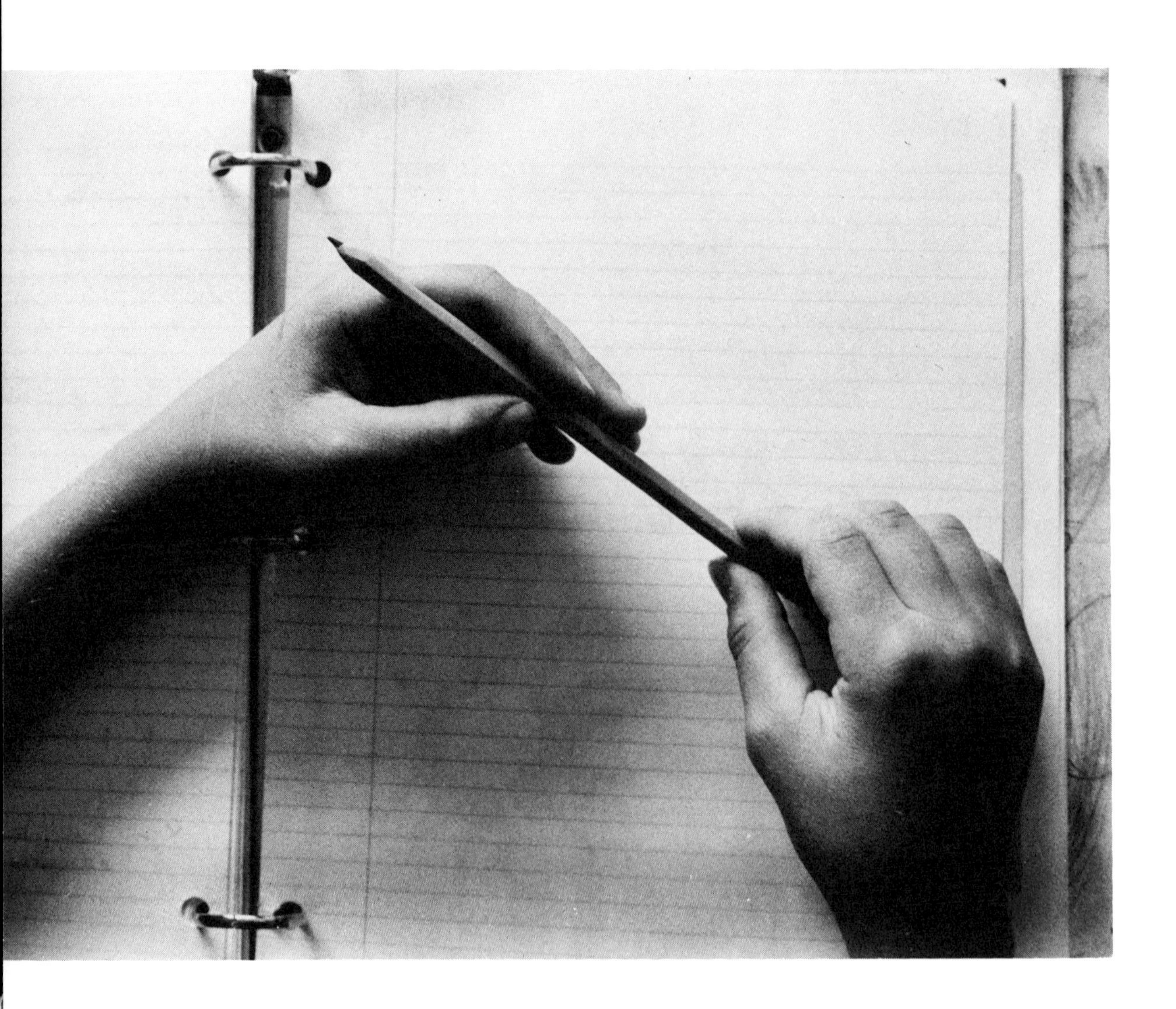

A while ago I became so curious about where and how things were made, and about the people who made them, that I decided to write a book about it. In case you're curious about who *I* am, here's a picture I took of myself in a mirror. And here's my daughter, Evan, who worked on the project with me. You'll see a lot of her.

Evan and I decided food would be a good subject to explore, because eating has always really interested us! In the supermarket, we began looking very carefully at the labels on the food we were buying. We found that the canned tomatoes we used in making spaghetti sauce came all the way from Italy. It was strange to think that these tomatoes had been grown on farms on the other side of the Atlantic Ocean, that they were raised and picked and

put in cans by people we'd never see. The cans traveled across the ocean in a freighter—something Evan and I have never done—and then they were trucked to our supermarket.

We wondered which of all the different kinds of food we ate came from farthest away. It turned out to be frozen lamb from New Zealand—that's thousands and thousands of miles from New England, where we live. Then we wondered what food was made closest to our own home. We figured produce that spoils quickly would have to come from nearby. What about fresh fruit and vegetables? The manager of our supermarket said that most of the produce we eat in New England in the winter comes from Florida or California. It travels by truck and train for several days to reach us.

We thought of milk, but because we lived in a large city, without farms nearby, we realized the dairies would be way out in the country.

How about bread? That was it! We found out that the bread we ate was made in a bakery just two miles from where we live.

6
BULKIE

WHITE
WHITE
WHITE
6
BULKIE
ROLLS

STAR
MARKETS

II.

AT THE BAKERY

The bakery turned out to be a three-story building with a huge flour storage silo on one side of it. The manager of the plant gave us permission to visit, and Abram Blanken, the quality-control manager, showed us around. He explained that it was his job to see that the baked goods looked and tasted just right. He also made up a lot of the recipes. And he understood why we were interested in the plant.

"When I was a boy in Holland," he told us, "I knew where all our food came from. I knew about the meat we ate; I saw the slaughterhouse and the butcher shop. My family were bakers, so of course I knew all about breadmaking. I knew where the grain—the wheat, you know—was milled into flour for our bread. Here it's different; here it all happens in factories."

Abram showed us the shop in the basement of the building for maintaining and repairing baking equipment. He took us to the dough-mixing room and to the business offices, with secretaries and management people, on the top floor. He told us that the baked goods were shaped and baked on the main floor. There we saw the huge ovens, the packaging machines, and the storage areas for the goods that were ready to be shipped to the supermarkets.

Abram explained that the bakery plant was open and running all day and all night. It never closed. At four in the morning, before sunrise, people were mixing and measuring, shaping and baking. They were doing it at eight o'clock, and at midday, and in the afternoon. They worked all through the evening, and even during the night when we were asleep. Of course, he said, the same people didn't work for twenty-four hours straight; each group worked

for eight hours, and then another group came to take their place. And the workers didn't only make bread, not at all. They made bread and rolls, cookies and cakes, pies and pastries—more than a hundred different items altogether. There were blueberry muffins and lemon-filled tarts, fancy iced cakes and plain hot-dog rolls. They seemed to make everything we could think of in the line of baked goods, and a lot we'd never imagined!

On Tuesdays and Sundays the baking stopped, and the bakery workers stayed home. On those days, the cleanup crew scrubbed and sanitized the plant from top to bottom, and then the baking began all over again.

Abram took Evan and me to every part of the building, and by the time he had finished, we were ready to start exploring on our own.

Moving Along with the Pastry

We headed straight for the pastry line, where hundreds of delicious coffee cakes were made every morning. The word "line" is short for assembly line. In factories, many different things are made by assembly lines—from cars to radios to pastry! The workers stand in a certain order in one place, and whatever it is that's being made passes by them on a moving conveyor belt. The belt is like that in a supermarket check-out line, only much longer. Each person does one part of the work required to complete the product. Employers find this a profitable way to have things made, but for workers, it is often very boring, since they repeat one small action over and over again all day long.

A huge batch of dough for the pastry line had been mixed and allowed to rise earlier in the morning. Then it had been divided up and rolled out into long thin sheets, spread with soft margarine, folded over, and rolled out again. This had been done many times, until there were seventy-two paper-thin layers of pastry in each sheet, and then each sheet had been rolled up once more.

Now it was time to make coffee cakes, so the machines of the pastry line were switched on. The long conveyor belt started moving; the large shining steel rollers began to turn smoothly; and the blades of the pastry cutter clicked swiftly up and down, up and down. The people who worked on this line hurried to their places, while the first man fed one end of a roll of pastry between the steel wheels to flatten it out still more. As the pastry passed through the machine, he quickly brushed the end of the roll with water, and stuck on the beginning of the next roll, so that there would be a continuous sheet of pastry yards and yards long.

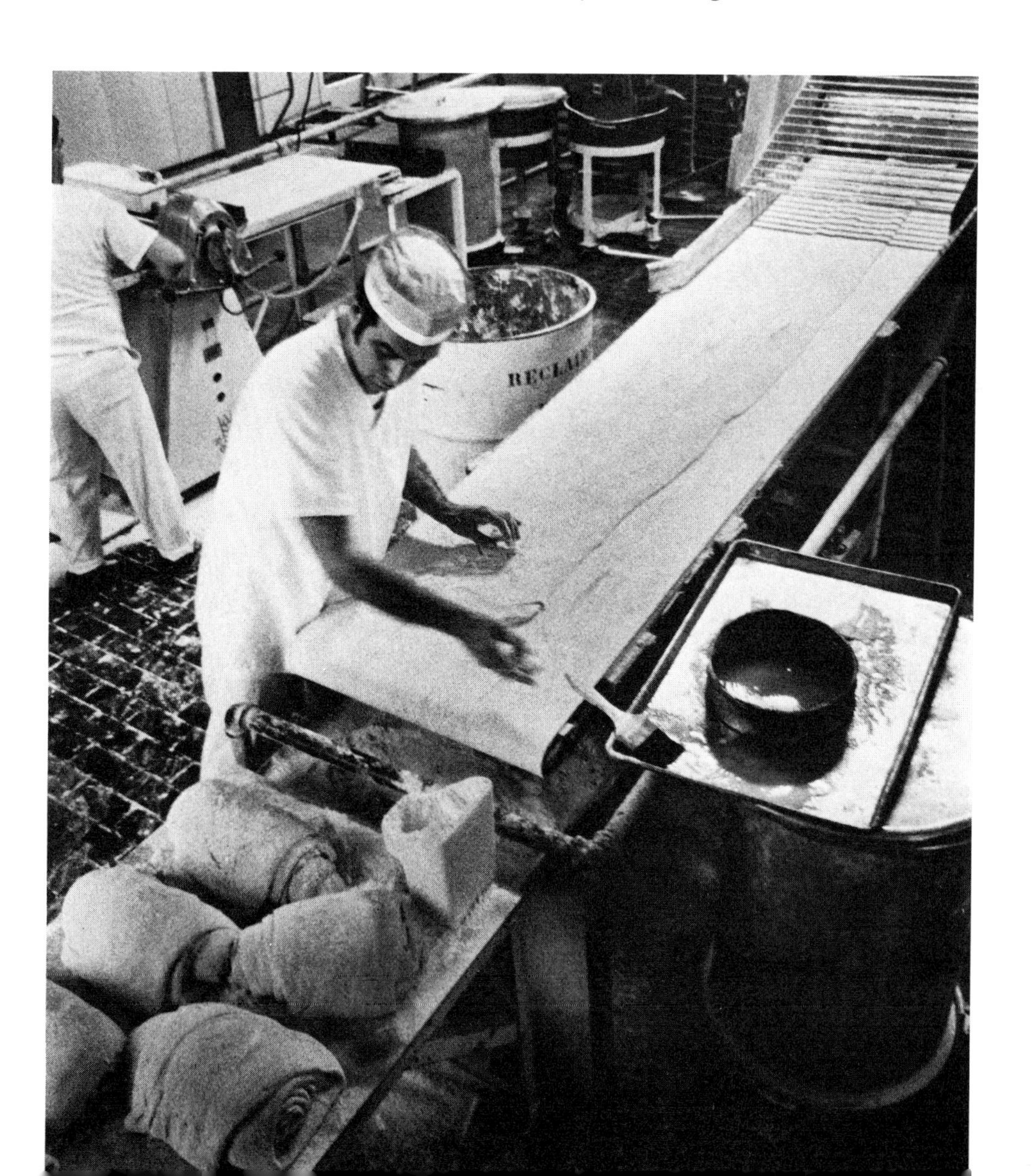

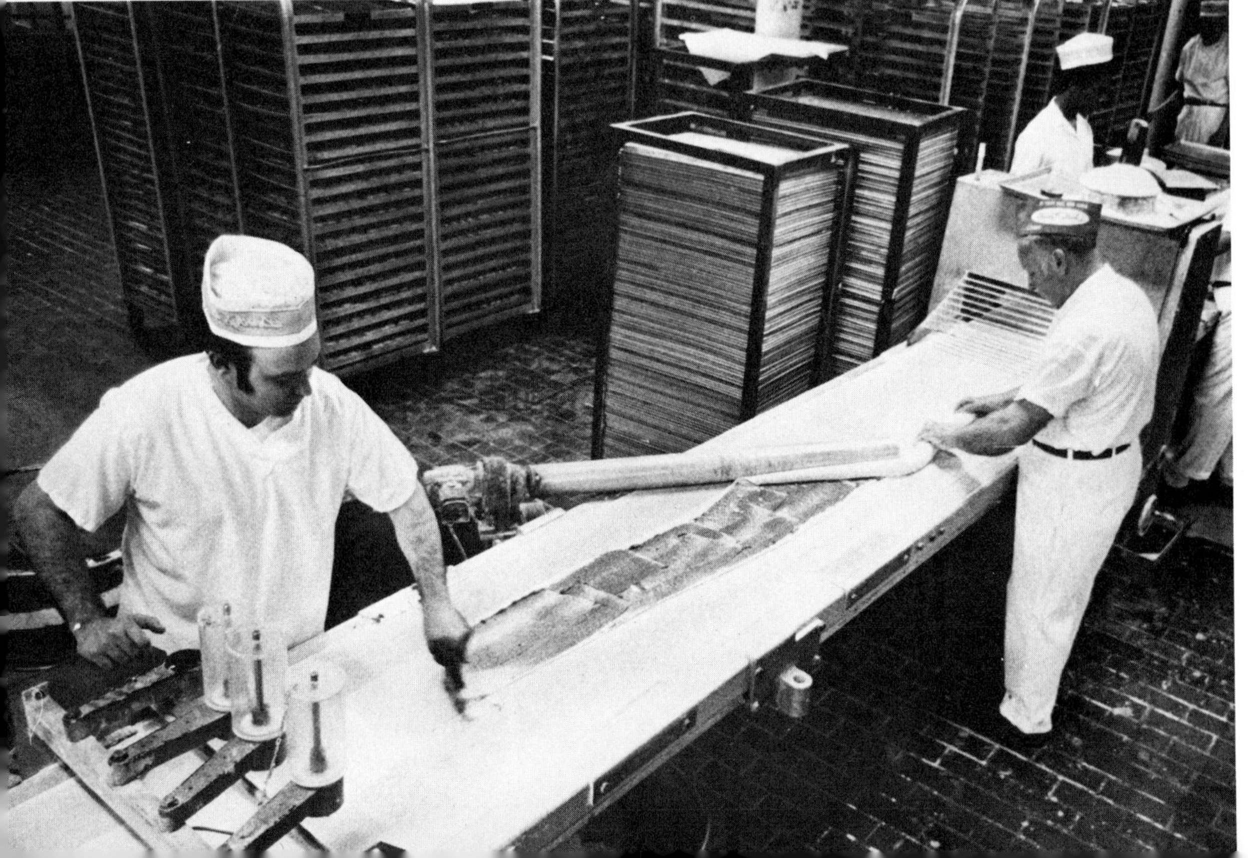

On the other side of the rollers, the sheet was spread with butter, brown sugar, and raisins. Then it bumped into a long metal roller lying on the belt and began to curl up. Here Evan and I stopped to talk to Bill Foley, the foreman of the line. He smiled and said it was fun to have us visit the plant. We asked him how long he'd worked in the bakery.

"I've been with the company for forty-two years," he said. "I've raised six children, and put two of them through college in that time. When I started work, I was a kid; I made eleven fifty a week."

"Forty-two years!" Evan exclaimed. "That's a long time—you've worked here that long?"

Bill laughed. "I've seen a lot of changes over the years. This used to be a small place, and we didn't have all this equipment on the line. We did most of the work by hand, but we're still a quality bakery, no mistake about that."

Evan watched the coil of pastry pass through a cutting machine where fast-moving blades sliced it into small rolls. The man on the other side of the cutting machine let her take his place on the line for a few minutes. She took an aluminum pan from a stack and put it on the moving belt where the next worker filled it with the rolls.

"Is this what you do?" she asked. "Am I doing it right?"

"Sure," he answered. "That's all I do."

At first she thought it was fun, but she soon grew tired of it. She lifted a pan off the stack, placed it on the belt. And another. And another. She began to make the gesture without thinking much about it, and the tins blurred in front of her. Her legs ached a little, and she shifted her weight from one foot to another, wishing she had a chair to sit on.

"I'd get tired of doing this all day long," she said. "Don't you?"

"Sure we do," he said. "That's the trouble with factory work. It's monotonous. But we make a lot of different kinds of things here, so we move around. When we finish work on the coffee cakes, we'll move to a different line and do the apple pies. Then we'll make hot-dog rolls."

The last man on the line put the pans of pastry on trays that he then slid onto racks in big wheeled carts, one after another, after another, after another. The filled carts were wheeled into the "proof box"—a sweet-smelling, warm, damp room where the yeast in the dough would make the pastry rise.

When the pastries had become puffed up with air to twice their original size, and the separate slices in each pan had become a single cake, they were loaded tray by tray onto moving shelves in a huge oven. The cakes passed through the oven like a slow freight train, and when they came out, twenty-five minutes later, they had been baked a golden brown and smelled wonderful.

While this batch of coffee cakes was cooling, we watched a machine ice some pastries that had been made earlier. As the pastries moved along on a belt, a machine pumped thin streams of white frosting over them. The whole machine jiggled constantly, so that the icing fell in patterns of curving, swirling, wiggly lines. Of course, we'd seen pastries like this in supermarkets, but we'd never imagined they were iced by such a terrific machine!

Meeting an Operator

The machine that wrapped up the pastries was part of a small, three-person assembly line. A man placed the pans of cooled pastry on a moving belt, where a piece of plastic automatically dropped over each one. A woman sat at the end of the belt, and as each tin dropped onto her lap, she quickly folded the plastic around it. She placed the wrapped cake on a warm conveyor above her, where the plastic melted just enough to become sealed around the tin. Then the tightly wrapped pastry moved along under a labeling machine that stamped it with a sticker, and a man loaded it into a basket at the end of the line.

The woman who folded the plastic, Bernadette Nicotra, was also the operator of this machine. She changed the labels when a different kind of pastry came along, and kept a careful record of how many of each product were labeled. As she sat wrapping package after package after package, she told us a lot about herself.

"I've been here for a long time, and I worked at other bakeries before this one. I've always worked; I believe you should use what you have. I have a seventy-five-year-old mother who'd *love* to be able to work right now. I'll tell any woman, go out and get a job. If you don't, married or not, you can't turn around and call a curtain your own!

"And I would say to any kid, if you want to get somewhere, go out and work. I had my first job when I was sixteen; I worked at an ice-cream place after school. I really wanted to be an airline stewardess; I wrote many a school report about it. But I missed the boat, I got married too young—I had to disregard it."

"Have you had many kinds of jobs?" I asked her.

"Oh, sure," she answered. "The first job I had after I got married was a typist's job. But the pay was lousy—the factories paid more than that. The next place I worked was in a factory where they made dolls, and my job was braiding the dolls' hair. I had all the different colored ribbons, and I put them in the hair. It was piecework, and if you made your quota, the money was good."

"What's piecework?" Evan asked.

"Well, you got paid so much for every doll you did. If you didn't do many dolls, you weren't paid much, but if you did a lot, the pay was pretty good."

"What was your next job?"

"Well, I needed regular work; at the doll place you got laid off a lot. I finally ended up working here, and I've been here eleven years now.

"You know, I feel some pride in myself. I've been able to bring up my three kids; I've been able to have an apartment and pay the rent, and I've been able to buy things for myself as I needed them. Independence is a wonderful thing. By the time I retire, when I'm old, I'll have a pension. Right now, I'm thinking about going back to school part-time. You like to think your time is spent useful, you know. If the bakery should close, I'd have something to fall back on. But I wouldn't choose to leave the bakery; I'm pretty satisfied right here."

Mary's Wonderful Wedding Cake

We found still another assembly line in the icing room. The first person on this line, a man, was knocking cakes out of pans and onto the belt with tremendously loud thumps. Then, as the cakes moved along, a woman set aside the broken ones, and several other women swiftly covered the good ones with a smooth, even layer of sweet icing. At the end, two people put the cakes into boxes.

The icers talked to each other over the loud rumble of the conveyor belt while they worked, but their eyes were always on the line of cakes passing constantly in front of them. They stood in their places, hands moving quickly back and forth between the chocolate cakes and the gigantic containers of yellow and white icing.

"Doesn't it make you hungry to ice cakes all day long?" we asked one of the women. "Don't you want to eat a lot?"

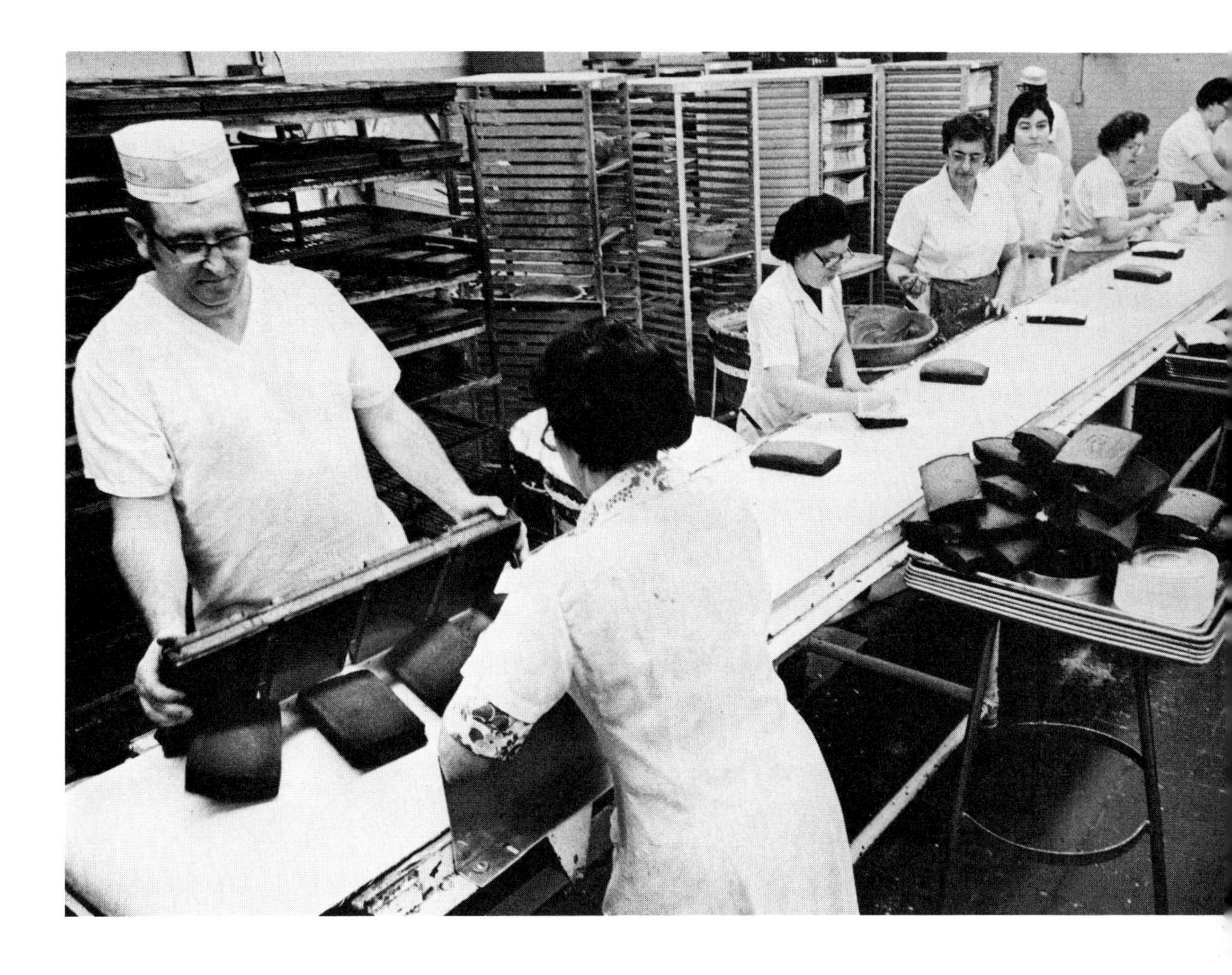

She laughed. "When you first start to work here, you do," she said. "The damaged baked goods are sent up to the lunchroom, and you can eat all you want. The first couple of days I worked here, I ate so much stuff I nearly got sick. Now I don't eat any of it!"

In back of the icing line we discovered the fancy cake decorators, Eric Plasis and Mary Evans. They decorated cakes for anniversaries, graduations, Mother's Day, Valentine's Day, and especially birthdays. And such birthday cakes! They made cakes with bunches of flowers on them, or ribbons and garlands, or even favorite TV characters.

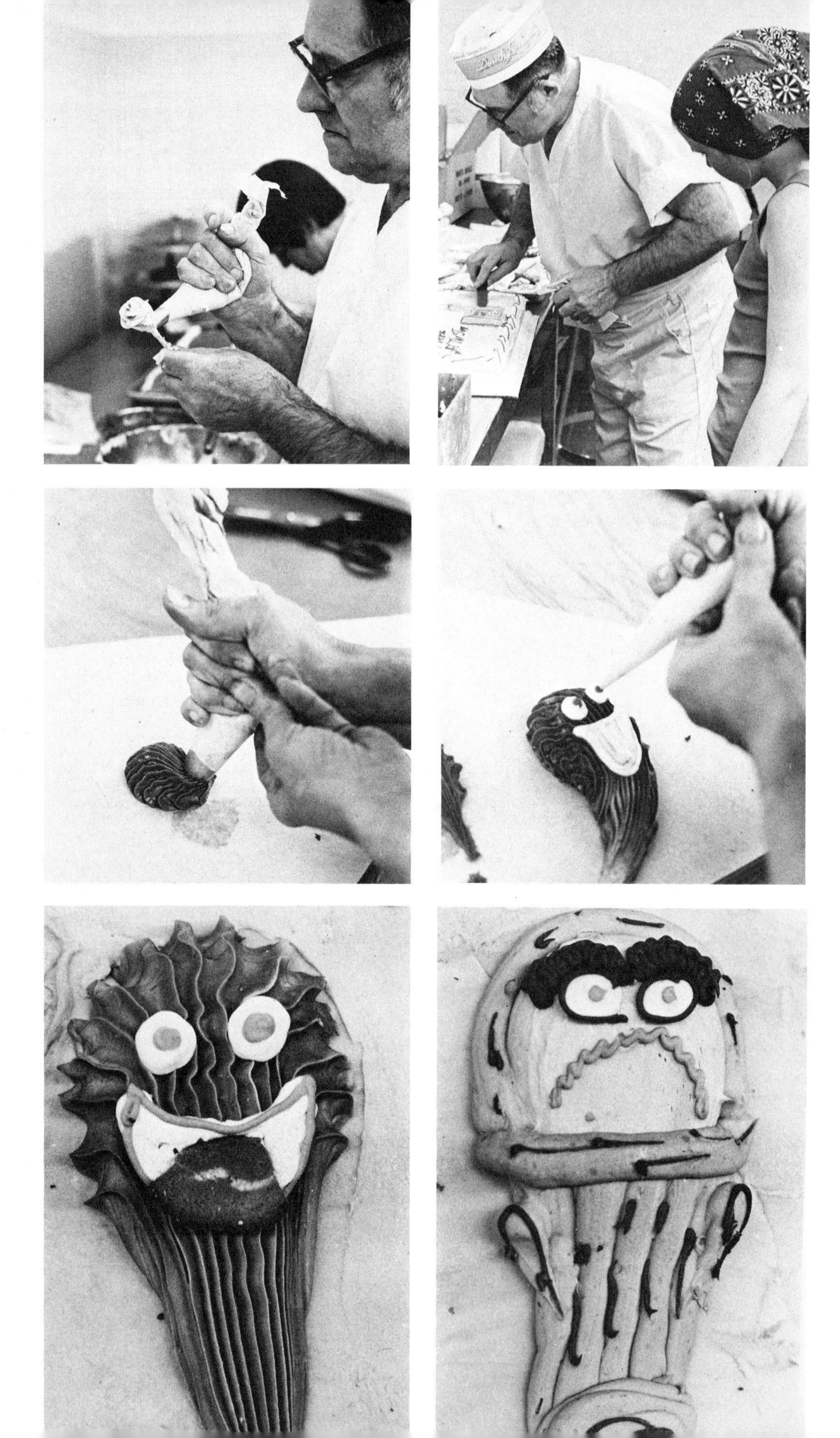

One day we watched Mary Evans decorate a huge wedding cake. It had been ordered by Albertine Callaghan, who worked in the icing line. Many years ago, when her granddaughter was small, Albertine had promised her a beautiful wedding cake decorated by Mary Evans. Now that the granddaughter was grown up and about to be married, Albertine had kept her promise. Mary had carefully chosen an icing design from one of her pattern books, and the day the cake was to be decorated, special sizes of layers were baked.

Mary had begun work first thing in the morning, for she knew the job would take many hours to complete. By noon she was half finished and ready to put on the delicate icing roses she had made earlier and set aside to harden. While Tom DiMambro, a worker from another part of the plant, proudly held the tray, she gently

put the roses in place, one by one, on the third layer, taking great care not to crush a single "petal."

Everyone in the plant knew that a special cake was being made that day, and many people dropped by the icing room to watch. When one of the secretaries from the business office came in, Tom began teasing her. "OK," he said, "when will it be your turn for one of these? How about it? You ready? You got a nice boyfriend?"

"Oh, get out of here, Tom," she answered, laughing.

Mary iced the next layer and carefully topped it with one of the platforms she had ordered from a cake-decorating company. Then she added little plastic figures of bridesmaids and groomsmen. Next she refilled one of her set of large metal-tipped tubes with soft white icing. Holding the cloth end tightly closed, she squeezed the tube and formed a delicately looped garland all around the top. She used a different tube to make small buds on the side of the cake, squeezing the cloth just enough to push out a small bit of icing. Done—and not a mistake on it.

She set the fourth layer in place, and started to work on the very last one—the layer with the figures of the bride and groom on top. The cake was more than three feet high now, but very steady. When the last decoration had flowed out of the tube, she set the layer gently in place, and the cake was finished!

The next day, after the cake had settled and the icing had set more firmly, Mary planned to separate the layers and put each one in a box. Albertine would take them all home, and the wonderful cake would be reassembled for the wedding.

Mary was proud of her skill, and of the fact that she had taught herself to be an expert cake decorator.

"There are schools where you can learn to do this," she said, "but I never went to one. I've been working here for twenty-three years. I've loaded the ovens and worked in the pastry line. When I was in the icing room, someone asked me if I'd like to try the fancy work. So that's what I've been doing ever since. And I like it. There's always something different to do. I keep making up new decorations and learning new patterns. I wouldn't want to work anywhere else now."

NOT EXIT

Not a Crumb in Sight

Once Evan and I went to the bakery on a cleaning day, and we hardly recognized it. The plant was almost deserted—only the office staff and the sanitation crew were there. The huge ovens were shut off, and all the moving, clattering, rumbling conveyor belts and other machines were still and silent. There was not a pie or a cake, a loaf of bread or a tin of pastry, anywhere in sight. The air was damp with clouds of steam, and all the floors were wet.

One man was using jets of hot water under great pressure to clean and sterilize the racks. A woman was scrubbing and mopping the walls and floors. Someone we couldn't see was even inside one of the huge ovens, scraping and cleaning it from top to bottom and from one end to the other.

By the time the crew was through, the whole plant was sparkling, and ready for the baking next day.

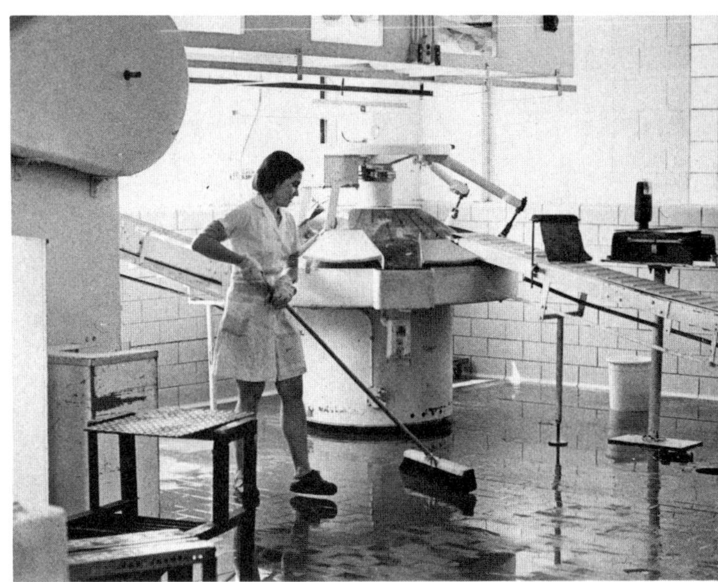

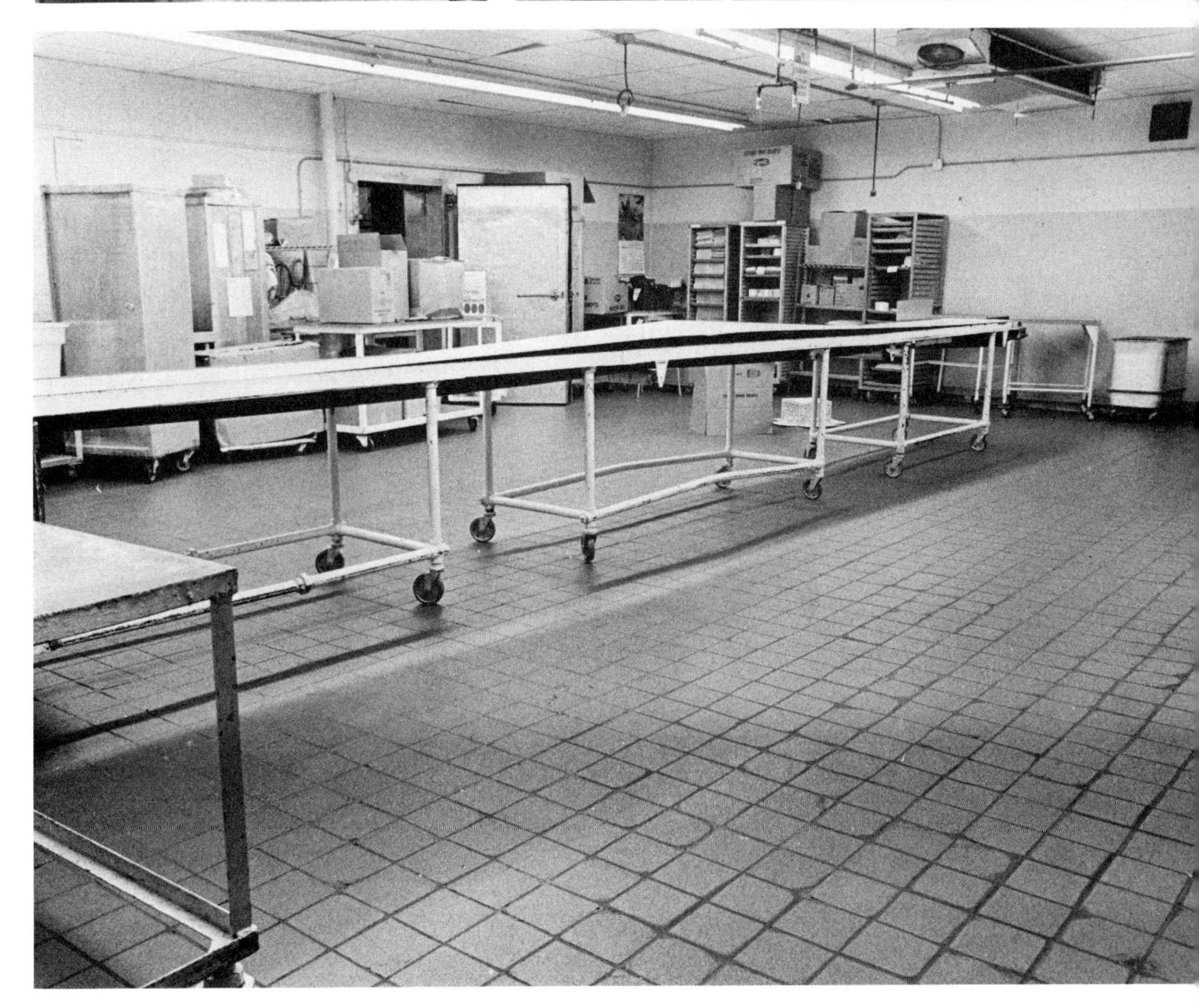

Bread by the Mile

Next we decided to follow the breadmaking process. It might not be as exciting to see as Mary's wonderful cake, but after all, it was the bread that had made us come to the bakery in the first place. We discovered that five mornings a week, a tank truck came to the plant, and the driver pumped flour from it into the huge, round, outdoor storage silo. The driver explained to us how this works.

"The flour has to be aerated—filled full of air—so that it will go through the hose and into the silo. There's an air compressor in the truck. It forces air into the flour, and then the flour just flows like water through the hose."

"How much flour are you pumping?" we asked.

"Well, let's see. This morning it'll be forty thousand pounds of flour. Sometimes it's as much as fifty thousand."

"Fifty thousand pounds!" we exclaimed. "Where does all this flour come from?"

7F
DRIVE
ALCO
EQUIPMENT
SAFELY
DRIVE
ALCO
EQUIPMENT
SAFELY

"That's a complicated question," he answered. "First of all, this flour is made entirely from wheat, but it isn't made from just one kind of wheat. I can't begin to tell you how many different kinds of wheat there are—there's high protein, low protein. There's spring wheat, and it's different from winter wheat.

"This flour is made mostly from wheat that's grown out west on the plains. The grain is loaded into trucks, freight cars, and barges, and carried to a mill in New York State. It's ground there, and mixed according to what customers order. The mill in New York State is where I came from early this morning."

"Do you make this same trip every day?" we asked.

"Oh, no. Every day's different. I deliver to a lot of bakeries. I might go to Rhode Island tomorrow, or Connecticut. I might start work at midnight or at four or seven A.M. There's no set pattern to it; that's how I live. Truck drivers are a different breed, you know. I've never worked in a factory a day in my life—I couldn't. The repetition would kill me. Now to drive a truck like this, you have to have a certain amount of competence. You don't get in a seventy-thousand-dollar truck and just race out. You've got to think you're the best, or there's no place for you in this business. I wouldn't do anything else. It's in my blood."

He knocked the last section of the tank truck with a hammer to make sure it was empty and went inside to fill out the delivery papers.

We headed into the plant to see how and where the flour came in. The mixing room was on the top floor, and here we discovered that the flour was pumped directly from the outdoor storage tank into a huge bread mixer at the back of the room. Water and syrup were piped straight in, too. We'd never thought about bread having sweetener in it, but most of the bread made here has some. Corn syrup was used instead of cane sugar, because it was less expensive and tasted just as good.

We watched the men turn some levers and wheels, and the exact amount of flour they wanted flowed into the mixer. We laughed when we saw the recipe they were following. It called for five hundred pounds of flour for one batch of bread!

"How much bread do you make every day?" we asked.

"About forty-five batches like this one."

"How many loaves does each batch make?"

"Seven hundred to eight hundred loaves," the men told us.

Evan and I were amazed. "That's . . . wait a minute . . . eight hundred times forty-five—that's *thirty-six thousand loaves of bread a day!*"

As we watched, dumbfounded, the men measured and added huge quantities of other ingredients—and then they closed the doors of the mixer and switched the machinery on. We heard the mixing arms inside begin to turn, and soon the hundreds of

pounds of dough were thumping around so loudly and heavily that the whole room seemed to shake. After about fifteen minutes, the men opened the door a bit to check its consistency. As the dough flew by, one of them quickly pulled out a small piece.

"It has to be just right—smooth and elastic," he said. "If the dough's mixed for either too long or too short a time, it won't keep its shape when it's made into loaves."

When the dough was just right, the men opened the door of the mixer wide. We saw the arms turning the huge batch of dough around and around, bumping it loudly against the sides of the mixer. The dough began to fall out, but was yanked inside again by the spinning arms. Finally it spilled and rolled, a gigantic flowing heap, out of the mixer and into the waiting container. The room seemed suddenly very quiet and still, as the men wheeled the dough into the next room.

5
1
12
6

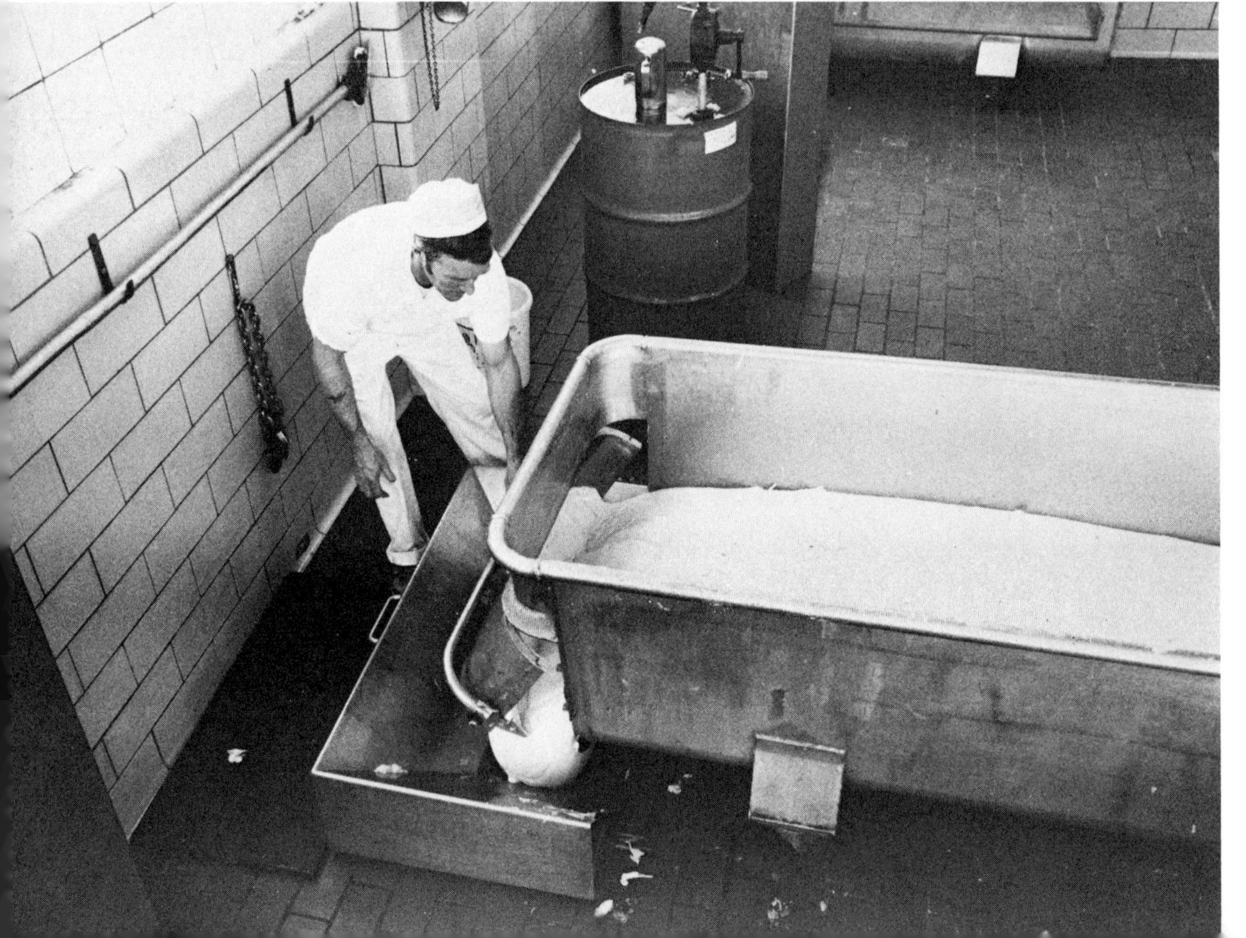

Here we talked with Frank Connolly, who'd worked in the plant longer than the others. We asked him how he liked his job.

"Well, I'll tell you," he said. "The hours are the most important thing to me. I work an early shift. That way I'm home when my kids come home from school. That's what's important to me, being home with my kids. And I like working up here. You're pretty much on your own in the mixing room. The pay's good, and the conditions are OK. I've worked all over the plant, and I prefer it here."

"What happens to the dough here," we asked.

"It will rise for three hours," Frank told us. "You can see from these other batches how much bigger this one will get. We'll mix in the remaining ingredients. The dough will rest for another fifteen minutes, then we'll send it downstairs. This batch here is ready to go. We're going to pour it right into this chute, which'll carry it downstairs to the divider."

We hurried down, and came into a room full of complicated machinery for shaping the dough into loaves of bread. Here a man was turning a wheel to adjust the divider that cut the dough into

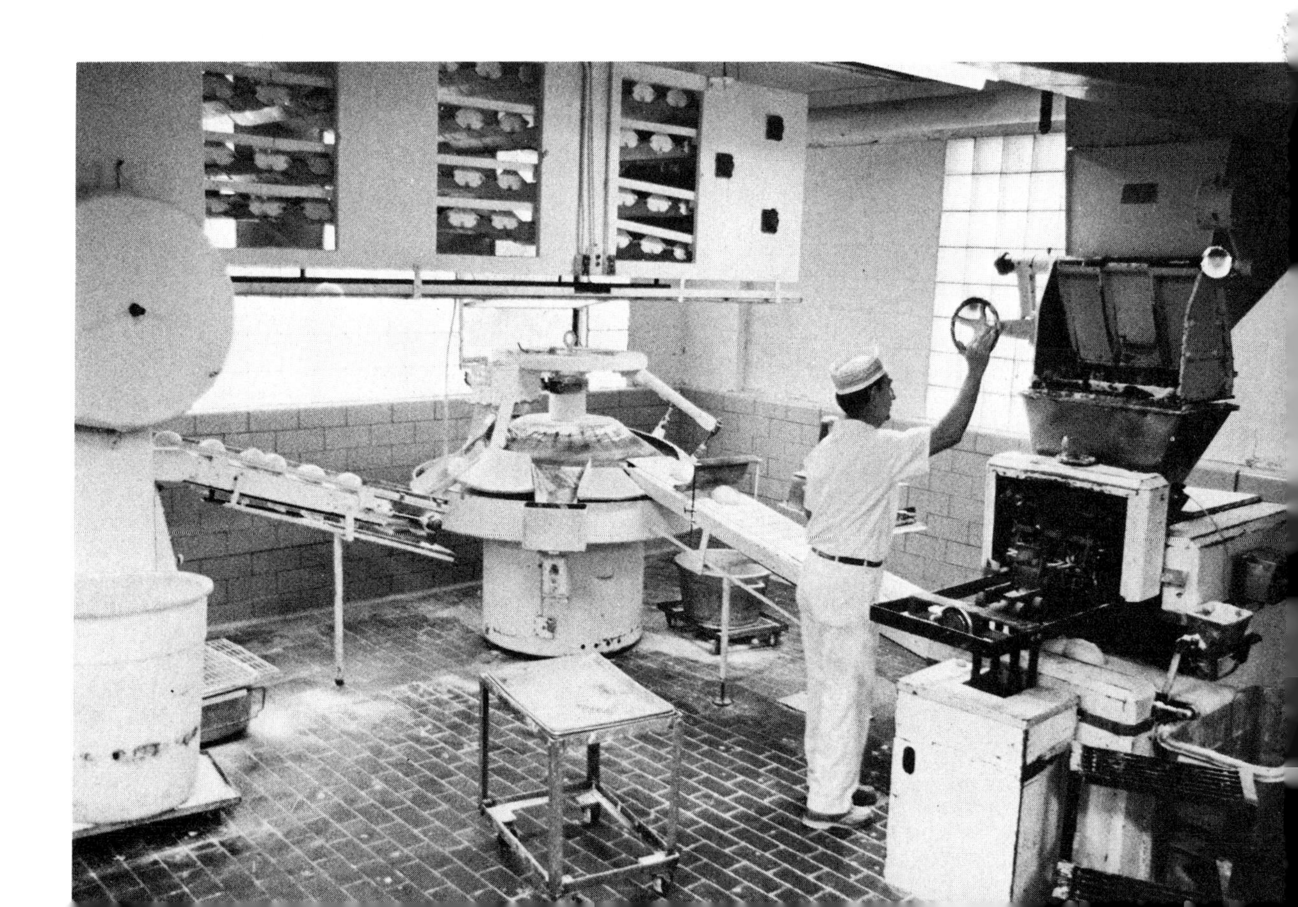

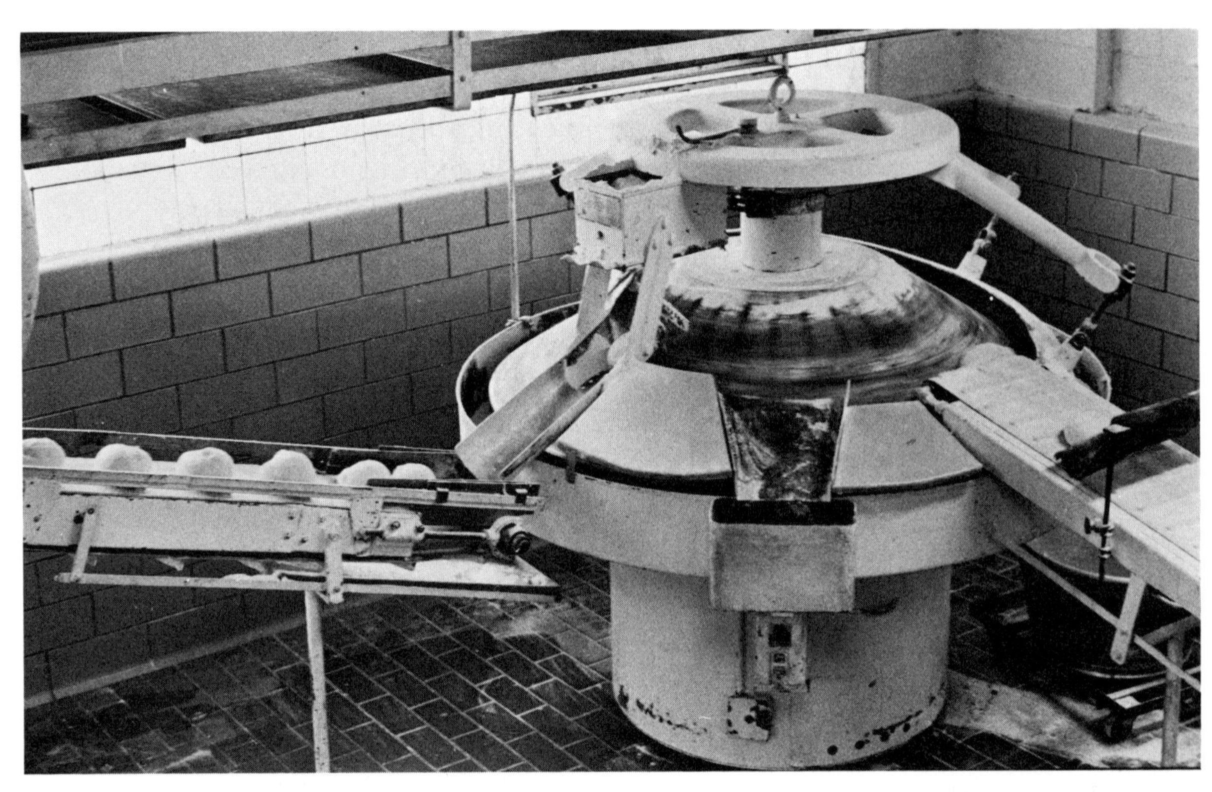

three loaf-sized hunks at a time. The pieces fell continuously onto the first of a series of conveyor belts that carried them on their journey around the room. As each piece moved along, it was flipped into a molder, whose spinning center whirled the dough around almost faster than we could see. This shaped the dough into a smooth ball, and then dropped it into a sling where it stayed, moving slowly along a track for eight minutes.

Next it was squeezed through rollers that flattened it out into a large disk, and then under a metal weight that caused it to roll up again. At last it dropped off the belt right into a pan, and this was the end of its journey for a while.

We watched a man load the pans onto carts and roll them into the nearby proof box for rising again. Another man was wheeling the carts of already risen dough to the oven, so we followed him.

We had never imagined an oven could be this large. It was even bigger than the oven in which the pastries were baked. It was the size of a small room. As we peered into it, we could see shelf after shelf of dough moving along past flaming gas jets.

As the sweat began to run down our faces, the man loading the pans into the oven grinned at us. "Wish we had air-conditioning?" he shouted over the squeak and rumble of the moving shelves.

"How do you stand it?" we asked.

"Oh, you get used to it," he answered. "But in summer it's wicked!"

"How much bread is baked at a time?" we asked.

"Depends on the size of the loaf. Could be as many as fourteen hundred loaves at once, when the oven's full. It works like this. I set the temperature—this bread is baked at four hundred and fifty degrees. Then I load the pans on these shelves, and they keep moving right on through the oven. I set the rate of speed of the shelves, too. It'll take twenty minutes for each loaf to go all the way through the oven—and then it'll be done. Look—here comes a batch out now."

Toasty brown, hot, and smelling delicious, pan after pan of bread rolled out of the oven and onto yet another conveyor belt.

"Follow the loaves along," he said, "and you'll see how they get cooled."

It was hard to see what the next machine did, because it enclosed the pans of bread pretty completely, but Evan and I laughed when we finally figured it out. Here hundreds of little, hollow, rubber suction cups on a roller lifted the loaves of bread up out of their pans by air pressure—like a giant vacuum cleaner. The empty pans then moved away on the conveyor belt, and the bread dropped onto another belt for its long cooling trip.

We had been told that each loaf of bread traveled on conveyor belts for almost half a mile around the plant until it was cool enough to be packed. As we began to follow the bread, we saw that the belt disappeared up through an opening in the ceiling. When we went upstairs, we came into a huge room filled with a maze of moving conveyor belt, all of it covered with freshly baked loaves of bread. We zigzagged around, trying to follow the belt to the end, and then found that it again disappeared—this time heading downstairs once more.

Downstairs we saw two men loading the bread into a machine that sliced a whole loaf at a time with one quick sweep of sharp blades. This was hard to see, since the blades were shielded so that no one could get cut on them. The bagging machine, next to the slicer, worked so quickly that we had a hard time figuring it out, until the operator showed us. As a jet of air blew up a plastic bag, long metal prongs slipped in and held the bag open. At the same time a loaf of bread was pushed right into the bag. Another device on the machine gathered up the open end of the bag and snapped a plastic tab around it. Then, as the package rolled around a corner, a label was slapped on top. At last, the loaf reached the end of

the belt. There men loaded the bread on to carts and wheeled it into the shipping room.

"Will this bread be sent to the stores right away?" we asked one of the men.

"No, not until evening," he said. "The trucks start going out around six o'clock. They're on the road all night."

Evan and I wanted to see how that worked too, so one evening we went back to the plant. We headed for the rear of the shipping room, squeezing past dozens of carts of freshly baked bread and pies, pastries and cakes. Here two huge trailer trucks were backed right up to the open doors of the loading platform, and men were wheeling carts of baked goods into the trucks.

We got to talking to one of the drivers about his job.

"I've worked here two years, and I like it," he said. "I come on at five and I work all night. I deliver all over. You know, there're

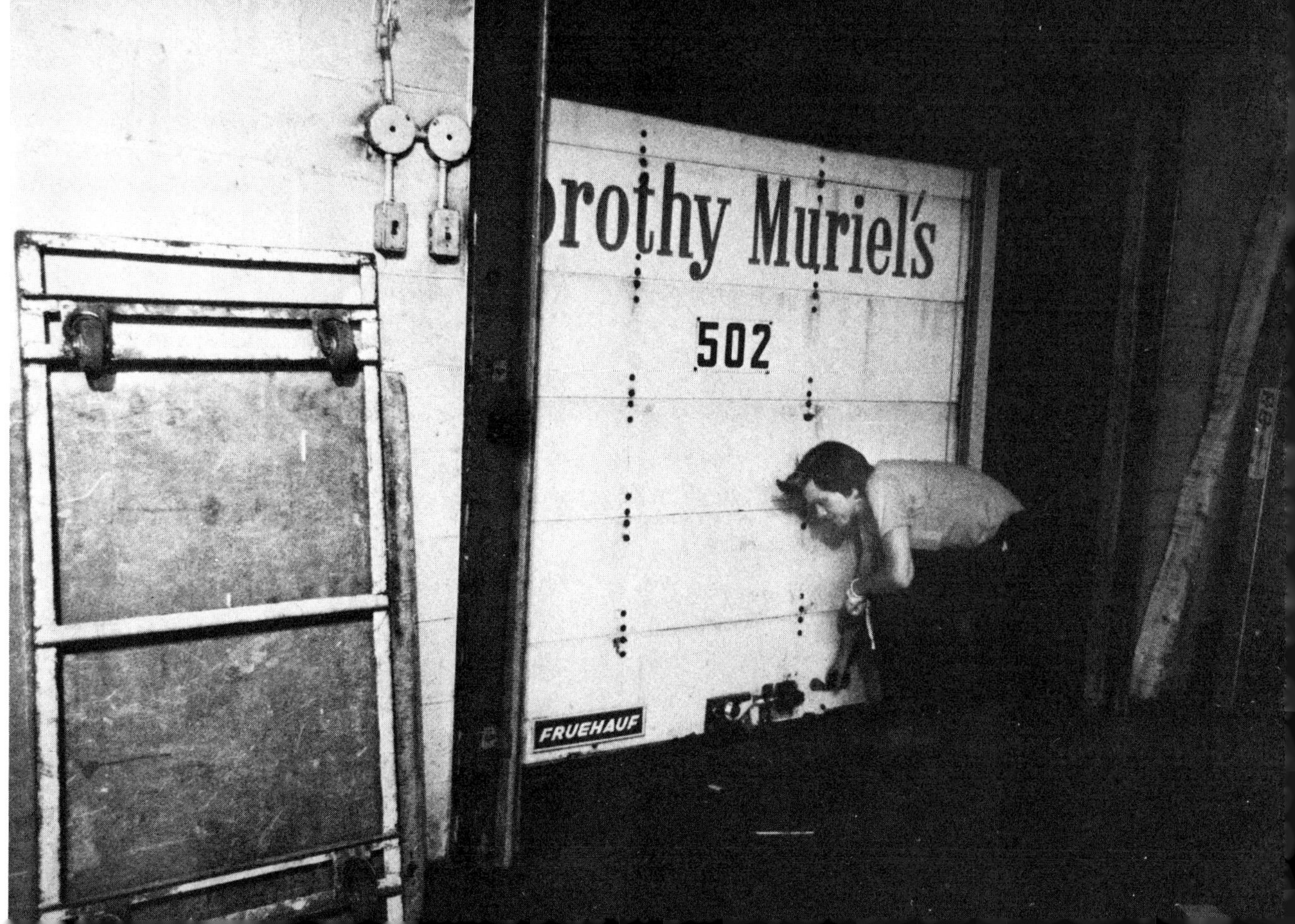
rothy Muriel's
502
FRUEHAUF

sixty-two stores we deliver to. Most of them are pretty near, but some of them are out of state. I might drive two hundred and fifty miles tonight."

"But how can you deliver stuff to supermarkets when they're closed?" Evan asked.

"Oh, I have a key to the delivery area of each store," he answered. "It's separate from the rest of the market."

"Is anyone there?" Evan asked.

"No, just me, unloading my truck. Then I lock up and go on to the next store."

We asked him how he got into trucking.

"I taught myself to drive," he said. "Some guys go to school, but I taught myself on my uncle's truck. It took me about three and a half weeks; then I went up for my class-one license. Someday I'd like to have a rig of my own. The trailer costs anywhere from twenty thousand dollars on up, but you don't necessarily have to own the trailer. Some companies have their own. Now the tractor—that's the cab and engine—that'll cost as much as forty thousand dollars. That's what I want; that's what I'm saving for—my own tractor. Maybe I'll be able to do it; maybe I won't."

"What's it like to work at night?" we asked.

"It's OK," he answered. "I really like driving at night, when there aren't all those inexperienced drivers on the road. Most people, they don't realize that when you're loaded behind, you can't stop quick like they can. They'll cut right in front of you, and you have to brake and pray!"

We watched while he finished loading his trailer and closed it up. As we started to drive home, we saw him pulling away from the plant. We were very careful not to cut in front of him!

"It's strange to think about his delivering all night, while we're asleep," I said. "I wouldn't like that job myself; I'd find it lonesome."

"I'd be scared, working in the supermarkets when it's dark," Evan said, "but he seems to like it."

At home that night, Evan and I started talking about the way people at the plant came to have their jobs and how they felt about them.

I said, "You know, a lot of the people we met at the plant seem fairly satisfied with their jobs, even though they find the work very monotonous. But others I talked to were really disappointed. They'd wanted a different career, but hadn't been able to follow it. They couldn't afford the right training, or jobs weren't available in the field that interested them. Some people didn't know *what* they wanted to do after they finished high school, and ended up as bakery workers almost by chance. Hang on. I'll read you some of my notes about what those people told me.

"One man said, 'I've worked here most of my life. When I got out of high school, I wanted to be a doctor, but I was an orphan, I lived in a foster home. There was no way I could get enough money to study to be a doctor.'

"Here's what another guy said: 'I'm not a baker; I'm an electrician. That's what I was studying in school, and I got good grades. But the building trade is shot right now, and I couldn't get an electrician's job.'

"And then another man told me this story. He said, 'I've been here fifteen years. I didn't plan on it. When I was in school, all I thought about was how soon I'd be through. I didn't think much about what kind of job I'd get after that. When I was still in school, I had an uncle who worked here. He got me a part-time job—I filled in for people who went on vacation. I thought I'd be finished at the end of the summer, but they wanted me to stay. So here I am, fifteen years later. At one time I got to thinking I wanted to go to college. I wanted to study science and math. I started out with a "prep" at night school; it was like a review of high school stuff. I did all right. Then I went into complicated math, and I couldn't do it. I guess I waited too long. I don't think I could do any other job now. I think it's too late.' "

Evan and I were quiet a minute, thinking about what these people had said. Then Evan remarked, "It's a real drag, not being able to do what you want to do."

"Why do you think those particular people keep working at the bakery?" I asked.

"Because the pay's good, and they have families to support, and that's that. And you've some security in a job like that; you'll get a pension when you retire, like Bernadette said."

"Do you think some of the people who work there *really enjoy* what they're doing?"

"Yes, I think so. The truck driver says he does. I can see he likes driving his truck when there's nobody on the road, and all that. Mary Evans says she does. She taught herself her job, and she's proud of it. People like what she does, too. I bet Abram Blanken likes his job. He's got a lot of authority, and he's doing what his family's done before him. I guess different people want different things in their jobs, and a few of them have been able to find what they want."

When Evan and I first went shopping after our visits to the plant, we headed straight for the baked goods. There were the usual racks of bread, rolls, and muffins, the same boxes of cookies and cupcakes that we'd always seen. But somehow they looked very different to us, now that we'd been to the bakery and talked with the people who worked there.

"Look," said Evan, "here are the coffee cakes. Remember the pastry line?"

"Sure," I answered. "Think how much we know about that pastry. We know that Abram worked out the recipe for it—"

Evan cut in. "And that Frank Connolly mixed the dough upstairs and dropped it through that hole in the floor—"

"And that Bill Foley curled it up in a roll—"

"And it was baked in the big oven—"

"And that Bernadette wrapped it—"

"And one of the truck drivers delivered it to this store late last night—"

"And then?" I asked.

"And then—" Evan answered.

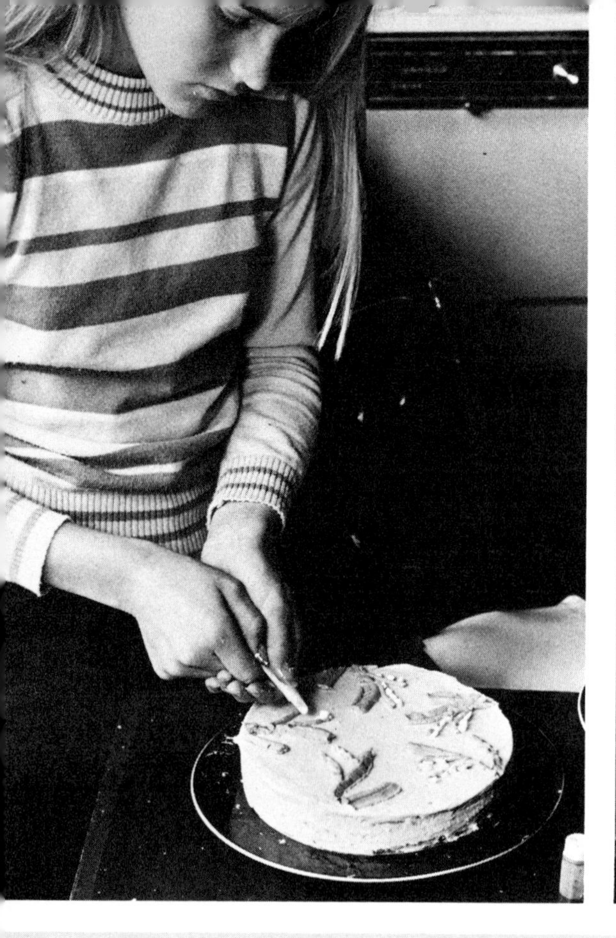

III.

TRY THIS

Where Did It Come From?

As we visited the bakery plant, we got ideas for some projects we decided it would be fun to try out with our family and friends. We thought you might be interested in them, too, so here they are.

We began by trying to learn where our food is actually produced. If you like, you can start out in the same way.

If you look in your refrigerator and on your kitchen shelves, you'll begin to discover that a lot of the food you eat has traveled a long way to reach you. You might want to get a map (or trace ours on page 57) and put pins on the places where the different foods come from. You'll have to do some detective work first.

Look carefully at the box and can labels to see what you can figure out. Many of them will say something like "Manufactured by Crumbly Crackers, Incorporated, Officeville, Illinois" or "Distributed by Fantastic Foods, Train City, New Jersey." That tells you

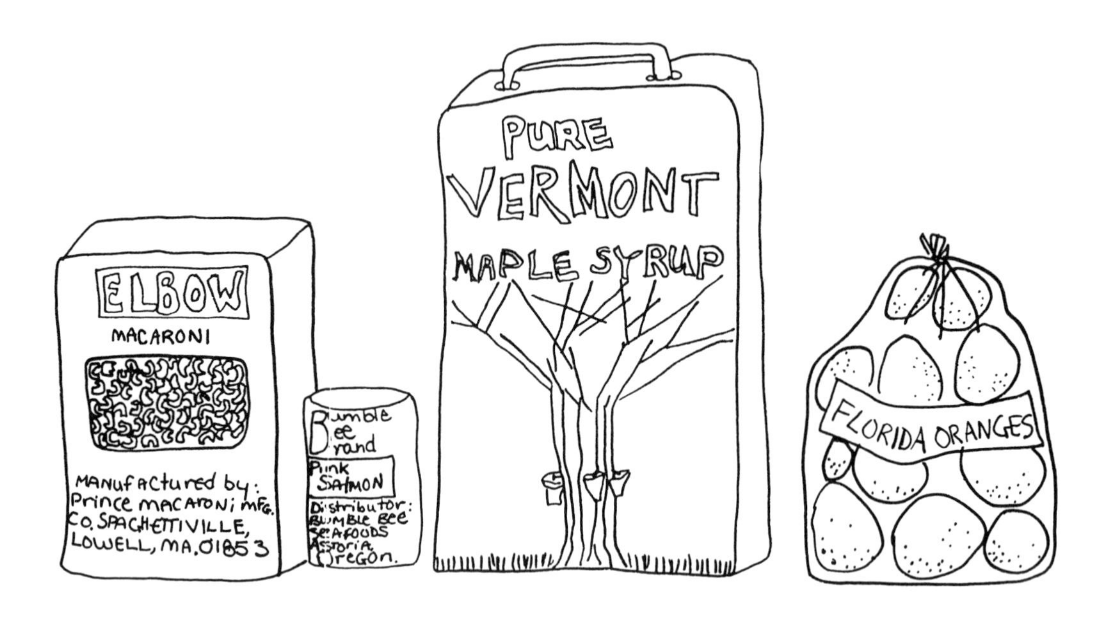

where the business offices of the companies are, but it doesn't tell you anything about where the food itself was grown and processed.

You're getting closer if you find a label that says something like "Packed by the Gigantic Vegetable Company, El Paso, Texas." Then you know that the food was canned in a factory in El Paso, even though it may have been raised in another part of Texas. And you've really tracked down your food if the label says "Product of." Such a food might be an imported one, like sardines from Norway, or fresh produce, like Florida oranges.

Another way to find out is to talk—either in person or on the phone—to the manager of the store where you shop. He knows a lot about the foods he sells, and the chances are he'll tell you things you hadn't even thought to ask, so don't feel shy.

When you've found out all you can, put the pins on your map and draw lines (or use string) from the pins to your home. You'll probably be surprised to see how many distant places your food comes from.

Here's a food map we made. Does yours look very different?

You can try this with other things in your home, too. The lines will probably be even longer! Was your television set made in Japan? Did any of your toys come from Hong Kong?

Suppose you had lived one hundred years ago in the same place you now live. What do you think your map would look like then? *Hint:* Think about airplanes, trucks, trains, boats, wagons, and horses. Think about what kinds of food can be raised in your area. If you live near a river, your map might have lines running along the waterways, because a lot of produce was carried by boat a hundred years ago. Do you suppose there were many farms near you? Or cattle ranches? Can you figure out what foods you'd have eaten in those days?

Here in New England, we would have had a lot of vegetables that grow in a short summer season and can be stored indoors through the snowy winter. We wouldn't have seen a head of lettuce or a tomato from November until June. Our map would have had a lot of very short lines on it—there weren't any planes to fly strawberries from California in those days!—and probably only a few very long lines. If we'd used pepper then, it might have been grown in Indonesia and been brought across the ocean in a sailing ship.

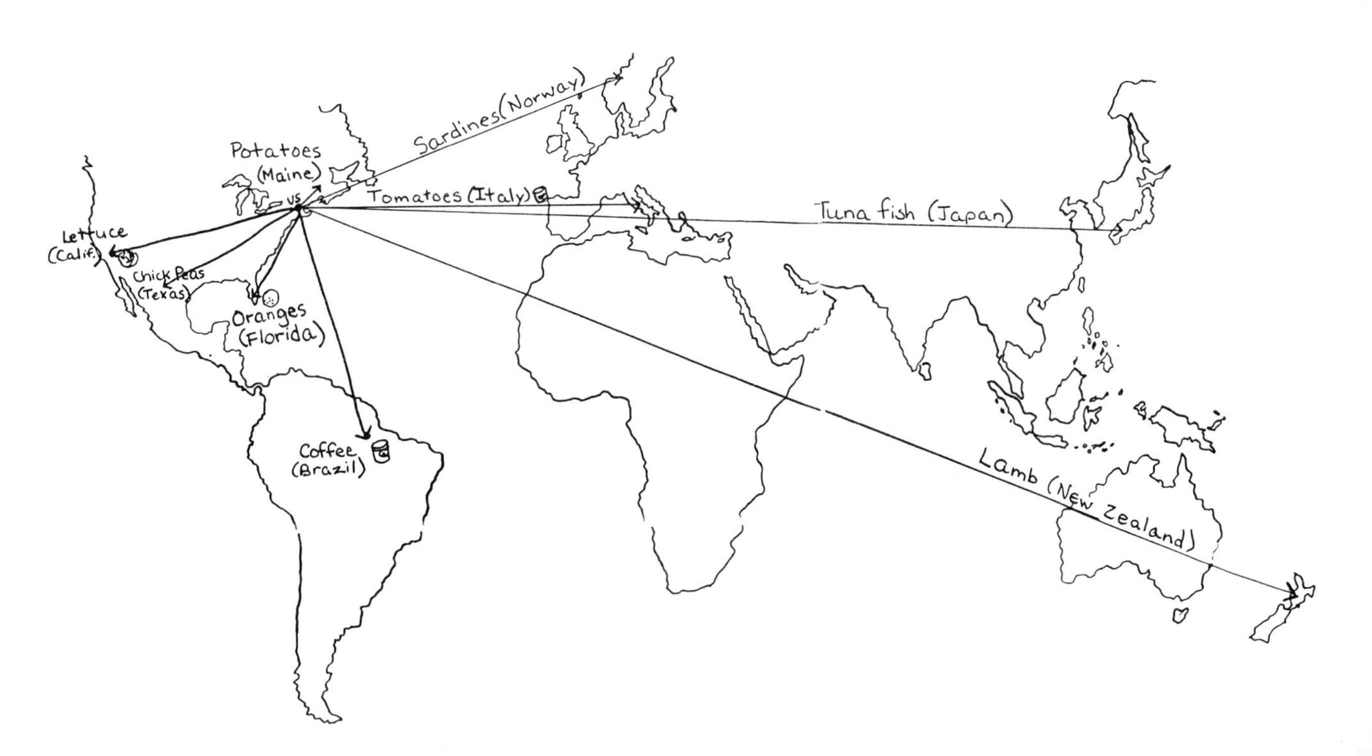

How's It Made?

Mass production (which means making a great quantity of something at one time) in a place like the bakery plant is quite different from making one loaf of bread or one coffee cake at home. But we still found we learned a lot from the two projects below—and our stomachs thought they were fine activities, too!

Baking Your Own

Abram took one of the bakery's gigantic bread recipes and reduced it so that we could make a couple of loaves at home. The loaves of bread were so delicious that we wished we could have mass-produced a hundred!

ABRAM BLANKEN'S BREAD

DISSOLVE

2 packages dry yeast

IN

2 $^{2}/_{3}$ cups warm water

ADD

2 tablespoons shortening

2 tablespoons honey

2 teaspoons salt

$^{1}/_{4}$ cup powdered milk

8 cups flour

Mix into a smooth dough, and let the dough rise in a warm place for 1 hour. Punch down the dough and knead it until smooth and elastic. Abram said, "Knead as long as you can!" (Ten minutes worked OK for us.)

Let the dough rise for 15 or 20 minutes.

Shape into two loaves and place in bread pans. Let rise for about an hour in a nice warm place. (75–80°).

Bake at 300° for about 30 minutes. After baking, brush with melted butter and cool on a rack. Eat and enjoy!

Here are some things to think about while you're kneading your dough:

How long will it take to make two loaves of bread?

How many loaves of bread do you suppose the bakery plant can make in the same length of time?

What do the answers to these questions tell you about machines and mass production?

Would it be cheaper for your family to make their own bread or to buy it in a supermarket?

Do people in your household like to bake bread, or would they rather do something else with their time, and buy bread ready-made?

You can ask these same questions about a lot of things in your home. Some people like to knit sweaters, build shelves, or bake pizza from scratch. Some people find they save money this way. Others buy most of their possessions. Which does your family do? Do you have friends or relatives who do things quite differently?

Mary Evans' Icing Devices

Mary uses commercial icing tubes for decorating her cakes as well as paper cones. She showed us how to make the cones, and we'll show you, but first you'll need to get together these materials: cake, any kind, covered with a plain frosting; some strong paper—heavy tracing paper is best, but strong typewriting paper will do; scissors; plus tape.

A NICE ICING

BEAT UNTIL FLUFFY
3 tablespoons shortening
SIFT AND BEAT IN
2 cups confectioners' sugar
ADD
1/4 teaspoon salt
2 teaspoons vanilla
If icing is too runny, add more
confectioners' sugar
If too stiff, add
milk, a teaspoon at a time.

The consistency of the icing is important: if it's too stiff, it will be hard to push out of the tube. If it's too runny, the decorations won't keep their shape. Experiment until you get it right. The recipe makes plenty of icing for decorations. Double it if you want to use the icing to frost your cake, too.

You might want to divide the icing up and tint it several colors—but add the food coloring slowly unless you want a very bright cake! Or you might want to leave all the icing white, as it is on Mary's wedding cake.

Now you are ready to make the three paper cones. Cut out two 8- by 8-inch squares of paper, fold each in half diagonally, and cut along the fold. Roll three of the triangles of paper into cones as shown. Put tape on the seams. Flatten the tips of the cones.

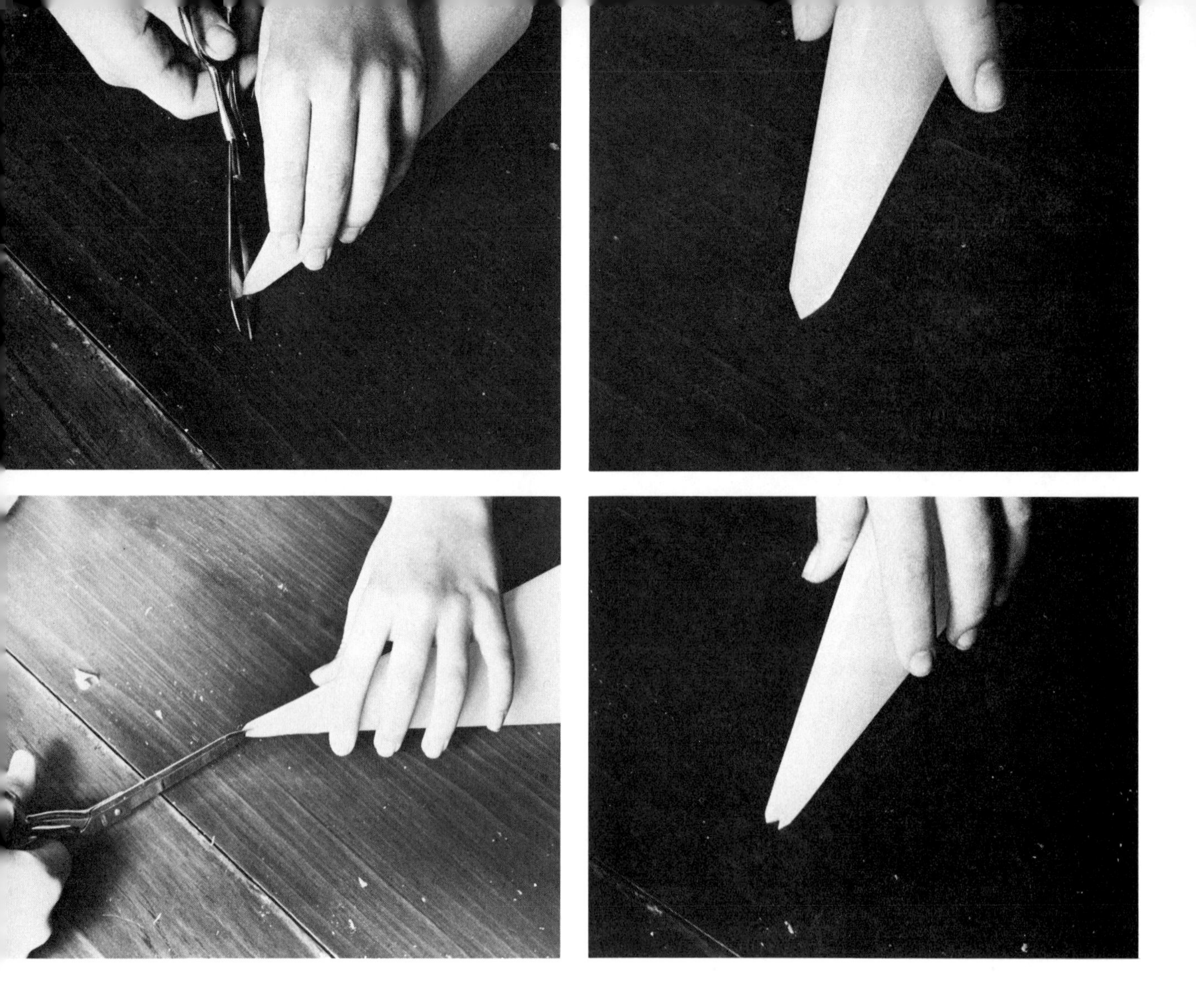

Cut straight across the first cone close to the tip. This will make a little round hole. Be sure to make all your cuts close to the tips—you can always enlarge the holes if you need to.

Cut the second cone as shown in the two photographs at the top of the page. Mary uses this shape cone for leaves and borders. You can think of new patterns to make with it.

Cut the third cone as shown in the two photographs above. Icing pushed out of this cone will look like flowers and rosettes. In fact you'll find it's good for all sorts of fancy designs.

Fill a cone with icing, and fold the paper over at the wide end to hold the icing in. You may want to practice your designs first on a plate. You can always scoop up the icing and put it back in the tube. Look at the pictures on pages 27 to 31, to get an idea of what Mary does, and then experiment on your own.

These are some of the cakes we decorated with family and friends: they're not much like Mary's, but we had a fine time making them.

We enjoyed learning how to do this, and we liked becoming more skillful at it. It made us feel proud of ourselves. We thought from what Mary said that she felt the same way. Are there things that you know how to do that make you happy? Are there things you'd *like* to know how to do? Are there ways you could learn about them?

Trips

After we learned how bread and cakes were made in a commercial bakery, we started to think about how other things we used were made. We visited some other workplaces—and if you'd like to do that too, we have some ideas you may find useful.

You might want to ask your parents or teachers to make the arrangements, but if you're up to doing it yourself, you can start by looking in the business pages of your phone book and picking out companies that make things you think are interesting. Be sure the companies are nearby.

This is how your phone conversation might go:

Receptionist (the person who answers the phone at the company): "Good morning, Dynamite Match Company."

You (try not to be nervous: remember, she's probably a nice person, but may be busy): "Hello, I'd like to know if kids can visit your factory."

Receptionist: "I'm sorry, but for safety reasons, company policy does not allow children under the age of fourteen into the plant."

(This is common. Don't be discouraged; just call another place.)

Receptionist: "Hello, Home Run Softballs, Incorporated."

You: "Hello, we'd like to visit your factory."

Receptionist: "Just a moment, please. I'll connect you with our public relations department." *R-r-r-r*

Public Relations: "Yes?"

You: "Hello, we'd like to visit your factory."

Public Relations: "Fine, we have tours for groups at scheduled times. . . ."

The public relations person will probably give you the information quickly, but if you're prepared, you can pick a time you and your friends, family, or classmates can go.

Almost any factory is interesting. Would you like to learn how potato chips are made? How your local newspaper is printed? How do you think shaving cream gets into spray cans?

Another way to visit workplaces is by exploring your neighborhood. It's important to be serious so that people won't think you're just fooling around. Simply walk into a store or factory or office that looks interesting, and say something like, "I'm studying different kinds of work, and I'd like to learn how you do things here." Lots of people will be interested to know that, and they'll take you all around. Don't be upset if some people show you the door. They may be afraid you'll steal from them, or slow them down, or get hurt; but you needn't let this make you feel badly.

Near our house we discovered a printing company where the machines that make the printing plates and the presses are really wonderful. We talked to the shoe repair person in our neighborhood, and watched him resole boots. He had learned his trade in Europe long ago, and liked having his own shop. Evan worked in a children's clothing store as a volunteer for a while, and learned about ordering, stocking, and selling goods.

What can you find in your neighborhood? A dry cleaners? A florist? A fish market?

Who Made It?

Almost everything you use was made by a person or a group of people. Who made this book you're holding? Some people made the paper from wood. Others made the ink from varnish and coloring. Still others ran the printing press and binding machine. I took the pictures for it and wrote the words. Sometimes writing books is very hard for me. I get ideas that turn out to be dumb, or I write badly. But when I do a book that I and other people think is good, I feel like the happiest person in the world.

Evan makes stuffed toys to sell. She does it because she wants to earn money, and because she enjoys sewing. She's good at it, too. When she's tired of sewing, she looks for other ways to earn money, because she doesn't like doing the same thing over and over.

Do you have a job that involves making something? Do your

parents? How do you, and they, feel about the work? You might find it interesting to ask your parents, grandparents, your neighbors, and friends about their jobs. They'll probably be eager to tell you, because work is so important to everyone. You might want to make a list of all the jobs your relatives have, or have had, and compare it with a list your friend makes. Are there a lot of different jobs on your list, or do the people you know tend to be in the same lines of work? Is your list very different from your friend's? See if you can find out what your great-grandparents and great-great-grandparents did. Are your parents or aunts and uncles doing the same work today?

Here's a list our friend Mark made.

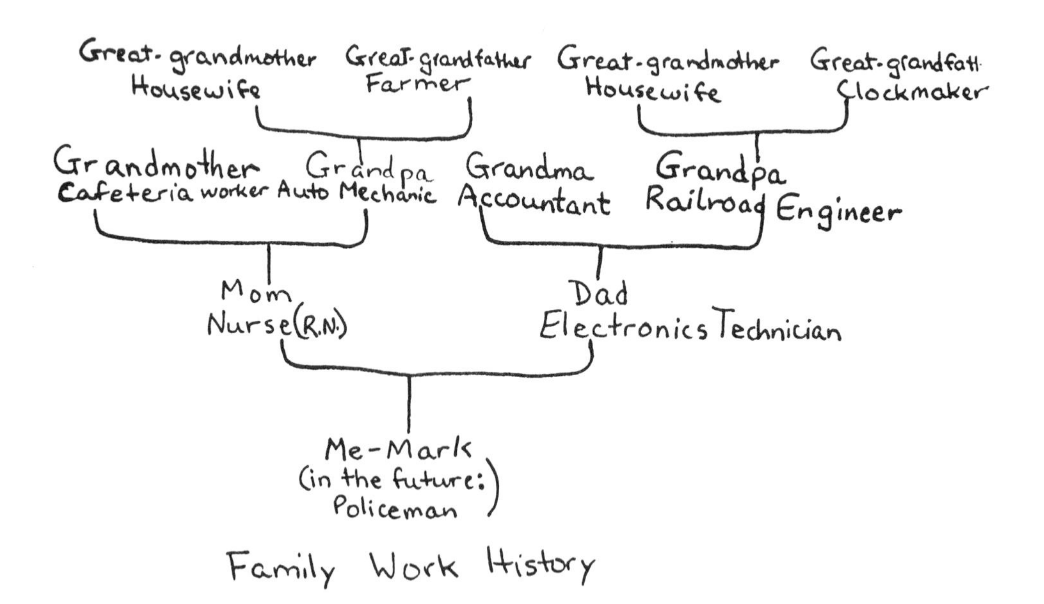

When you talk to people about their jobs, you might ask some of these questions:

Do they like and value the work? What is pleasing to them about it?

If they don't like the work, why are they still doing it?

How important is the amount of money they earn?

Are the hours they work important to them?

What about the conditions in the workplace—are they dangerous, or noisy, or ugly, or pleasant, or what?

How about their independence? Do they work for themselves or for someone else?

You can look through this book and think about how some of the people at the bakery might answer these questions. If you really become involved in the subject, you might like to write up your interviews, add photographs or drawings, and share the project with your classmates at school. You'll be well on the way to making a book like this one, and you may find it very interesting work.

We did.

Index

ABOUT THE AUTHOR

"In the United States," Aylette Jenness writes, "universal schooling and protective child-labor laws have ironically (and necessarily) deprived children of an important part of their education—knowledge about the workaday world, where in fact they will spend the greater part of their lives." That concern, and her own curiosity about where and how the things we use every day are made, led her to write this book, which, she hopes, "will stimulate children to think about their future place in the working world."

Aylette Jenness is the author of a number of distinguished photodocumentaries, including *Dwellers of the Tundra,* about life in an Alaskan Eskimo village, which was selected as an Honor Book in the 1970 Book World Children's Spring Book Festival; *Along the Niger River: An African Way of Life;* and, with Lisa W. Kroeber, *A Life of Their Own: An Indian Family in Latin America,* which in 1975 was chosen as an Honor Book by the New York Academy of Sciences.

Aylette Jenness and Company: *First row,* Matt, home cake decorator; Matthew, principal illustrator; Aylette; Mark, family tree specialist. *Second row,* Bonnie, friendly critic; Corinna, home cake decorator; Sam, trouble-shooter; Danny, friendly critic. *Third row,* Evan, Aylette's daughter; Eve, illustrator.

Photo by Kirik Jenness

REUBEN'S FIRST

Christmas CROW

Steven M. VanderWest

Illustrated by Renae Wallace

ISBN: 978-1-4497-5164-7 (sc)

Library of Congress Control Number: 2012908859

WestBow Press books may be ordered through booksellers or by contacting:

WestBow Press
A Division of Thomas Nelson
1663 Liberty Drive
Bloomington, IN 47403
www.westbowpress.com
1-(866) 928-1240

Printed in the United States of America

WestBow Press rev. date: 5/30/2012